S0-DUU-264

THE WORLD HERITAGE

CORAL REEFS

CHILDRENS PRESS®
CHICAGO

Table of Contents

Library of Congress Cataloging-in-Publication Data

Larramendi, Alberto Ruiz de.
[Arrecifes de coral. English]
Coral Reefs / by Alberto Ruiz de Larramendi.
p. cm. — (The World heritage)
Translation of: Arrecifes de Coral
Includes index.
Summary: Discusses the habits of coral polyps, the formation of various types of coral reefs, associated plant and animal life, and the challenge of conservation.
ISBN 0-516-08384-8
1. Coral reef biology—Juvenile literature. 2. Corals—Juvenile literature. 3. Coral reefs and islands—Juvenile literature. [1. Coral reef biology. 2. Corals. 3. Coral reefs and islands.]
QH95.8.R8513 1993
574.5'26367—dc20

93-3438
CIP
AC

ISBN (UNESCO) 92-3-102601-1
ISBN (Childrens Press) 0-516-08384-8

Coral Reefs

It's fascinating to watch underwater scenes of coral reefs and the spectacular creatures that live there. It's even more intriguing when you learn some of the secrets of this natural world beneath the ocean's surface.

People once believed that coral islands were formed by some geological process—the same way that mountains and continents were formed. Actually, coral islands and reefs result from the buildup of millions of limestone shells left by tiny animals called polyps.

What coral polyps have done over millions of years is astounding. In constructing reefs—and entire islands—they have surpassed all the efforts of architects throughout human history.

The Living Coral
Though they look like stone, coral reefs and atolls are the work of tiny animals called polyps. Polyps form an external limestone skeleton to protect their soft internal parts. Over millions of years, living polyps pile up on the skeletons of those who came before, creating coral formations such as the one in the photo on the left. On the right is a sea anemone *(top)* and a butterfly fish *(bottom)*.

The Underwater Kingdom of Coral

"We passed through the Low or Dangerous Archipelago, and saw several of those most curious rings of coral land, just rising above the water's edge, which have been called Lagoon Islands. A long and brilliantly-white beach is capped by a margin of green vegetation; and the strip, looking either way, rapidly narrows away in the distance, and sinks beneath the horizon. From the mast-head a wide expanse of smooth water can be seen within the ring. These low hollow coral islands bear no proportion to the vast ocean out of which they abruptly rise; and it seems wonderful, that such weak invaders are not overwhelmed by the all-powerful and never-tiring waves of that great sea miscalled the Pacific."

—Shipboard diary of Charles Darwin
(*Voyage of the Beagle*, 1835)

Charles Darwin is best known as the father of the theory of biological evolution. According to this theory, animal and plant species over time undergo a process called natural selection. This idea is sometimes shortened to simply "survival of the fittest." That is, creatures that adapt best to their environment—gradually changing in response to the changing conditions around them—are the ones that survive.

But the theory of evolution was not Darwin's only contribution to nineteenth-century science. In 1842, this eminent English scientist published *Structure and Distribution of Coral Reefs*. This book is the basis of our present knowledge of these fascinating formations.

Today coral reefs are found in nearly all of the earth's warm and temperate seas. They do not grow well in temperatures below 68 degrees Fahrenheit (20 degrees Celsius). Most of the earth's warm ocean currents skirt along the eastern coasts of the continents, while cooler currents move along the continents' western edges. Thus, coral reefs are more abundant along eastern coasts. Examples of this pattern are the reefs of the Indian Ocean, east of the African continent; the Caribbean reefs, east of North and Central America; the Polynesian reefs, east of the Asian continent; and the Great Barrier Reef, along the east coast of Australia.

Oceanographers calculate that the earth's oldest coral reefs are at least 350 million years old. At the time they arose, there were coral reefs throughout nearly all of the world's oceans.

Close to the Surface
Coral colonies usually grow close to the ocean's surface. There is such a variety of coral species that they create the appearance of an undersea garden. This photo shows a colony of *Acropora*.

The oceans have changed drastically since then. But the remains of ancient coral can still be found in places such as the Harz Mountains of Germany and the Tyrol region of Austria. Both of these areas, far from any ocean today, were once covered by seas.

There are three distinct types of coral reefs, classified according to their structure. Fringing reefs are supported by a rocky substratum and usually run along a coastline. Barrier reefs are some distance from the shore and are partly submerged in the water. An atoll is a circular reef surrounding a lagoon.

Aldabra Atoll is a perfect example of the third category. On the other hand, Australia's Great Barrier Reef is made up of all three types. This makes it the world's finest example of coral architecture.

The Aldabra Atoll

Between India and the great island of Madagascar, the Indian Ocean is dotted with thousands of tiny islands and reefs. Because of their poor soil and lack of fresh water, most of these islands have no human inhabitants.

Plants or Animals?
The classic differences between plant and animal life are not always clear in the world of coral. Despite its branch-like appearance, the sea plume *(left)* belongs to one of the most primitive classes of animals, the crinoids. It is always seen attached to elevated places on the ocean floor, where it has easy access to food carried by the currents. To the right is a starfish.

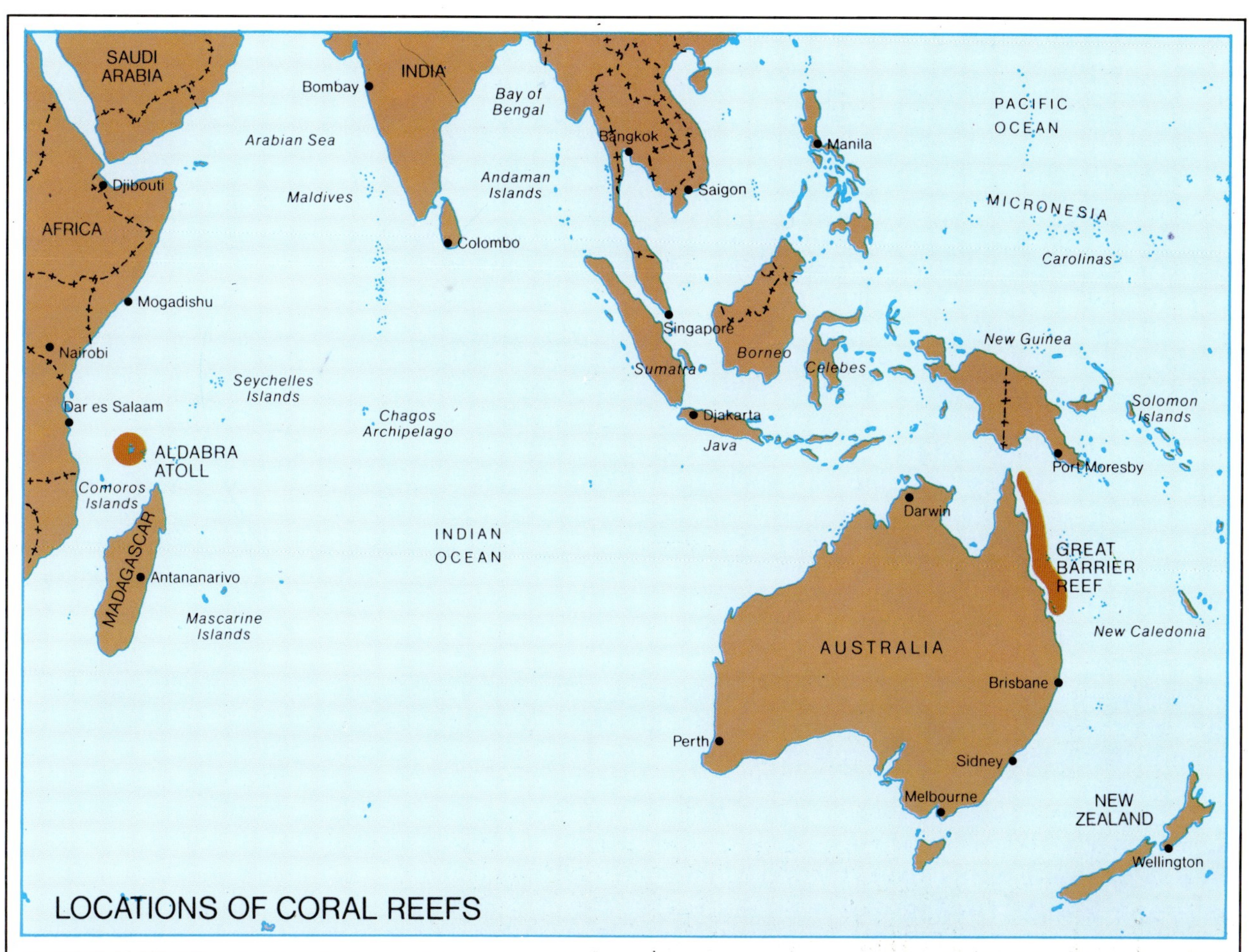
SAUDI ARABIA
Bombay
INDIA
Bay of Bengal
Arabian Sea
Djibouti
AFRICA
Maldives
Andaman Islands
Bangkok
Saigon
Manila
PACIFIC OCEAN
MICRONESIA
Carolinas
Colombo
Mogadishu
Nairobi
Singapore
Borneo
Sumatra
Celebes
New Guinea
Seychelles Islands
Dar es Salaam
Chagos Archipelago
Djakarta
Java
Solomon Islands
ALDABRA ATOLL
Comoros Islands
Port Moresby
Darwin
INDIAN OCEAN
MADAGASCAR
Antananarivo
GREAT BARRIER REEF
Mascarine Islands
New Caledonia
AUSTRALIA
Brisbane
Perth
Sidney
Melbourne
NEW ZEALAND
Wellington
LOCATIONS OF CORAL REEFS

Nevertheless, these islands have an exceptional climate. It is semi-arid, with an intense rainy season lasting from November to April. With such a climate, and the singular beauty of the land and sea, these islets fit the image of a tropical paradise. This is a perfect description of Aldabra Atoll.

Geographically, Aldabra is directly opposite the coast of Tanzania, 250 miles (400 kilometers) northwest of Madagascar. Politically, it belongs to the Seychelles Islands.

The atoll consists of four main islets of coraline limestone, separated by narrow channels. The islets enclose a shallow lagoon. Nowhere does the land rise more than 100 feet (30 meters) above sea level. Yet the surface is varied and rough, due to its coral origin and the erosion from the waves.

The islands' inner coast, facing the lagoon, is adorned with imposing mangrove swamps. These swamps extend out into the warm, shallow waters of the lagoon. Dotting the coast are fine, sandy beaches. The outer coast is the "active" part of the coral reef. The atoll is entirely surrounded by an outer reef, which protects it from the pounding of the sea.

The total surface of Aldabra is 135 square miles (350 square kilometers), of which only about half is dry land. The rest is composed of mangrove swamps, beaches, the outer reef, and the inner lagoon.

According to Darwin, coral reefs rest on ancient volcanic islands that gradually sank far below the surface of the sea. In the first phase in the evolution of an atoll, a fringe of coral forms around the edge of the sunken island. In later stages, the coral fringe grows upward and outward, while the island continues to sink. This forms the inner lagoon.

Timeline

1726	The work of the French surgeon Jean Andre Peysonnel shows that the polyps which form coral are animals.
1769	Captain Cook, commander of the vessel *Endeavour*, discovers the Great Barrier Reef of Australia, adding it to contemporary maps.
1971	The Seychelles declare Aldabra Atoll a special reserve.
1975	The Great Barrier Reef of Australia is declared a national park.
1981	UNESCO includes the Great Barrier Reef as a natural resource on the World Heritage list.
1982	UNESCO adds Aldabra Atoll to the World Heritage list.

The Biggest Mollusk in the World

The tridacna, a giant clam, is the largest shelled mollusk on earth. It weighs over 450 pounds (200 kilograms), and its valves are more than 3 feet (1 meter) long. It's hard to imagine that our familiar little clams are relatives of this giant species. A one-celled alga lives in its tissues, just as algae live in the tissues of coral polyps. The tridacna's shell is used as a holy water font in many churches.

The Magic of Color

The living things that occupy coral reefs have adjusted to their unique environment in many ways. The fishes that swim around the reef display remarkably dazzling colors. In some cases, these brilliant colors help the fish to mark out their territory. In other cases, the coloration serves to confuse predators.

In the final phase, the volcanic island disappears completely, leaving a lagoon enclosed by a necklace of coral reefs. The lagoon is shallowest in the center, above the last remains of the sunken island. The complete process may take as long as a million years.

The most outstanding animal life on Aldabra is a colony of giant turtles, about 152,000 in number. In India, this species of turtle was driven to extinction, as its meat was highly valued by sailors. On Aldabra, it reaches densities of up to 70 turtles per acre (170 per hectare). Aldabra is the only place on the planet where the dominant herbivore, or plant-eater, is a reptile.

The Aldabra Atoll also provides refuge for Carey turtles and green turtles. It is estimated that one thousand female green turtles come to the coast every year to lay their eggs.

The atoll has important colonies of seabirds, too, such as frigate birds and red-footed gannets. But Aldabra's true bird stars are a whitethroat species and a drongo species that cannot be found anywhere else on earth.

Aldabra's terrestrial marsh hens have a population estimated at five thousand. This is the only species of marsh hen in the Indian Ocean region to escape destruction by humans.

There is one element that scientists cannot measure when they judge a place's value: beauty. Aside from their sheer beauty, the uniqueness and harmony of these islands make the Aldabra Atoll one of the most stunning natural landscapes on the planet.

The Mystery of Coral

Until the eighteenth century, scientists pondered whether the polyps that made up coral were animals or plants. In 1706, the Italian naturalist Luigi Marsigli claimed to have proof that coral was "a true plant, that has milky sap in its bark and produces flowers and fruit."

The mystery was finally solved in 1726, thanks to the work of the French surgeon Jean Andre Peysonnel and the Englishman John Ellis. These scientists showed beyond a doubt that polyps were carnivorous animals that fed on ocean plankton. Today we classify the polyp as an animal of the phylum Cnidaria. Some other animals in this phylum are jellyfish, hydras, and sea anemones.

Flightless Birds

In some island environments, the birds have no land predators. These birds have lost the ability to fly. This is the case with several species of marsh hen throughout the world. These birds can defend themselves only by running or hiding. They are especially in danger from some of the animals that humans have introduced—animals that can hunt them. That is why all the species of marsh hen in India are now extinct. Aldabra's land marsh hen *(lower right)* is still flourishing, with a population of about 5,000 individuals. Below is a photo of the Aldabra Atoll. At the right is a fish whose dull, gray coloration helps it blend in with its coral habitat.

The Voyage of Captain Cook

In the spring of 1768, the British Admiralty and the Royal Society of Science in London were in a commotion over a singular event expected to take place in June of the following year: the passing of the planet Venus between the sun and the earth. At this time navigation relied on the clock and the sextant, so astronomy was vitally important. The occasion would be an opportunity for scientists to confirm their calculations of the distance between earth and the center of the solar system.

For this reason, a long voyage was planned by eminent mathematicians and astronomers. Their objective: to study Venus from an observatory in Tahiti, an island south of the equator in the Pacific Ocean. The vessel chosen was the *Endeavour,* a 366-ton coal tender, built in the shipyards of John Walker. Commanding the *Endeavour* was Captain James Cook, a seaman well versed in southern crossings. The departure date of the expedition was set for August 25, 1768, from the British port of Plymouth.

Once the path of Venus was observed, Captain Cook opened the secret instructions that were given to him by the British Admiralty. He was to search for a yet unexplored continent called "Terra Australis Incognita." This land was thought to form part of New Zealand, Tasmania, and New Guinea. The new voyage took Cook to New Zealand. Carefully he mapped its coasts, observing the natural history of the islands. Later he reached Australia at a point that members of the expedition named "Botany Bay," not far from present-day Sydney.

On the night of June 11, 1769, after landing at several points along the Australian coast, the *Endeavour* ran violently aground on a coral reef. The crew immediately lowered the sails and launched a boat to find out how serious the accident might be. The ship was stuck fast, and three major leaks threatened to keep it from floating again. Cook ordered the men to bail. Cannons, iron and stone ballast, kegs, flasks of oil, and partially spoiled provisions were thrown overboard in a desperate attempt to reduce excess weight. But it was not enough. The men envisioned a tragic resolution to their predicament. As a last resort, Captain Cook hauled down the masts and rigging, and put the ship under oars. At last success smiled on the expedition. After twelve exhausting hours, the *Endeavour* was afloat once more, soon reaching deep water.

This was Captain Cook's dramatic discovery of Australia's Great Barrier Reef in 1769. To this day, the point where his ship ran aground is known as Endeavour Reef, while the section of coast where the boat was repaired is called the Cape of Tribulations.

Since Captain Cook's near tragedy, more than 500 ships have been lost on the jagged reefs of coral. Some are famous, such as the frigate *Pandora* that went down in 1791, carrying fourteen sailors from the mutiny on the *Bounty* from Tahiti to stand trial in London. There was Matthew Flinders' corvette, the transatlantic *Yongala,* that ran aground with some 22 passengers after being battered by a cyclone. Captain Flinders had meticulously charted the reef and warned other captains to stay away from it if they hadn't the courage to "thread the needle...amongst the reefs." His own ship went down on Wreck Reef in 1803.

Light and Food, Sustenance for Coral

The shape of a coral's surface depends on the ocean currents over the reef, the amount of light that penetrates the water, and the way each coral species tends to form colonies. Some coral is primarily vertical, while other forms have a large horizontal surface. Among the horizontal types is a species called elk horn coral *(right).*

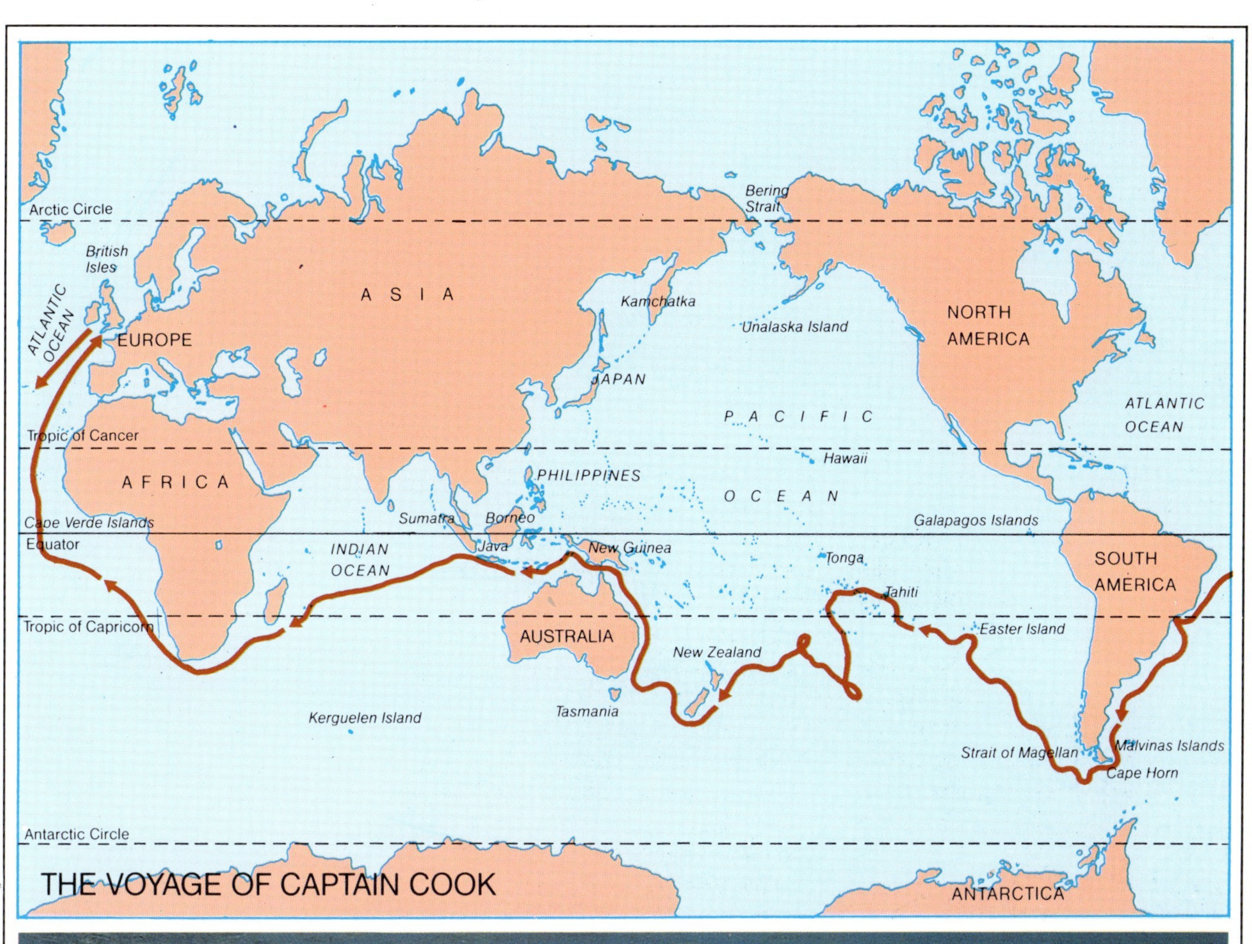
THE VOYAGE OF CAPTAIN COOK
Arctic Circle
British Isles
ATLANTIC OCEAN
EUROPE
ASIA
Bering Strait
Kamchatka
Unalaska Island
NORTH AMERICA
JAPAN
PACIFIC
ATLANTIC OCEAN
Tropic of Cancer
Hawaii
AFRICA
PHILIPPINES
OCEAN
Cape Verde Islands
Sumatra
Borneo
Galapagos Islands
Equator
INDIAN OCEAN
Java
New Guinea
Tonga
SOUTH AMERICA
Tahiti
Tropic of Capricorn
AUSTRALIA
Easter Island
New Zealand
Tasmania
Kerguelen Island
Strait of Magellan
Malvinas Islands
Cape Horn
Antarctic Circle
ANTARCTICA

Barely half an inch (1.5 centimeters) in length, coral polyps are pouch-like in form. These primitive creatures consist simply of a double wall of tissue, within which the digestive processes take place. The organism has a crown of tentacles around its mouth. As these tentacles wave, they stir up a gentle water current that brings food within eating range. Polyps can also inject a stinging substance into their prey.

The polyp secretes a substance that contains calcium. That is how it forms a protective outer skeleton. Millions of these skeletons, piled on top of each other, result in a coral reef.

Polyp species have been on earth for millions of years. One of the factors in their successful survival is the way they reproduce. Polyps use both sexual and asexual reproductive techniques.

To establish a new reef, they use the sexual process. The male expels a cloud of spermatozoa into the ocean, fertilizing the ova of the female. The result is a swarm of tiny, pear-shaped larvae—the planulae. Each planula has a fringe of cilia, or hairlike projections, that enable it to swim. Eventually the larvae fasten themselves onto a rocky surface, where they later develop into adult polyps.

Once the pioneer colony is firmly settled, it turns to the asexual reproductive method. The polyps form buds and branches, which become new polyps. These, in turn, branch into still more descendants. This method is much faster and more energy-efficient than sexual reproduction, allowing the colony to grow rapidly. Then the limestone skeletons of dead polyps begin to accumulate.

Two Hundred Million Years Old

A reminder of the remote epochs of the earth's formation, turtles originated during the Triassic Period, about 200 million years ago. Aldabra Atoll serves as refuge for a population of 152,000 giant turtles *(lower right)*. Here they reach densities of up to 44,000 individuals per square mile (17,000 per square kilometer). These islands are among the few places in the world where the most abundant herbivore is a reptile. Some of Aldabra's interior lagoons are shown in the upper right photo. At the left is a tridacna, or giant clam.

Occupations on the Reef

Most of the organisms that make up the structure of the reef take in food by filtering tiny particles that are suspended in sea water. But this work is done in a variety of ways. The polyps of certain solid coral forms are nocturnal (active at night), helping them to stay safe from predators. The *Gorgonia*, or sea fan *(right)* sends out a fine net of filaments that can filter a great volume of water. It is not surprising that other animals that feed by filtration also use the coral colony as their base. Some, such as the crinoid, crawl here and there along the reef.

Coral takes a great variety of forms, depending on the depth of the water, the strength and direction of currents, and the orientation of the polyp colonies. Rapidly growing species occupy the upper levels. They form broad, leaflike expanses and have advantages in competing for light. In contrast, the lower parts of the reef are dominated by rounder forms, hard and solid. Although their growth is slow—only one-fifth of an inch (half a centimeter) a year—these are the true architects of the reef.

Various other material acts as a "filler," strengthening the reef's structure. It is held together with lime (a calcium compound) secreted by certain animals and plants. An army of invertebrates, boring into the coral to find shelter, creates fine sediment. Larger fragments include the skeletons and shells of various sea creatures and chunks of coral broken off by the waves. All these materials are held together by lime-secreting algae. The algae's branches hold calcium carbonate in place, forming a kind of cement that seals cracks in the reef.

Zooxanthellae— Coral's Inseparable Companion

It has long been known that living coral is not found more than 130 to 165 feet (40 to 50 meters) deep, depending on the clarity of the water. Below this point, living polyps disappear, and the reef is composed entirely of limestone skeletons.

Based on this fact, a team of scientists placed living polyps in dark boxes. The water had the correct temperature and salinity, and the polyps received all the proper nutrients. But the coral grew only one-tenth as much as it would in lighter conditions.

Since coral polyps are tiny animals, why is light necessary for their growth? Zooxanthellae offer the answer.

Zooxanthellae are tiny, one-celled algae found in the polyps' tissues. They are found in concentrations as high as 30,000 algal cells per cubic millimeter. (That would be almost 500 million per cubic inch.) Light is essential for their survival and growth.

The role of these plants in the growth of coral is still not clearly understood. At first, it was believed that zooxanthellae transferred oxygen to the polyps.

A Pearl in the Sea
This beautiful photo shows a classic example of a coral atoll. A relatively deep interior lagoon is enclosed by a ring of coral that has been colonized by vegetation. Beaten by the waves, the upper part of the atoll is rough and craggy. Beaches form in areas protected from the thrusts of the sea, made of fragments of the reef. The ocean makes a magnificent background for the atoll when viewed from above.

Today the most widely accepted theory is that the algae help the coral by consuming the coral's waste products. During photosynthesis, the algae give off carbon, hydrogen, and nitrogen that the polyps can use. The polyps give off nitrogen and phosphorus that the algae can use.

More importantly, in producing limestone (calcium carbonate), the polyp releases carbon dioxide that the algae need for photosynthesis. By removing carbon dioxide from the scene, the algae help the polyps in their skeleton-building. Here is a case of symbiosis between a polyp and an alga, in which both organisms benefit. Because algae need light to photosynthesize, the reef must have light in order to grow.

The Great Barrier Reef

Without a doubt, the most remarkable example of the tiny polyp's work is the Great Barrier Reef. It covers some 100,000 square miles (260,000 square kilometers)—an area about half the size of Spain—along Australia's eastern coast. More than 250 species of coral live here in countless colonies.

The Great Barrier Reef forms an irregular arc that begins near the Murray Islands and the Torres Strait and ends near the Capricorn Archipelago, north of Brisbane. Measuring about 1,250 miles (2,000 kilometers) long, it is replete with reefs, cays, islands, and islets. It is the largest, most spectacular creation by living beings on the planet.

A Living Forest
The undersea world challenges our idea of what is an animal and what is a plant. Here we see strange living things that look like plants in shape and color. But in reality most of them belong to the animal kingdom. The photos show: a mushroom coral *(left)*, a *dendronefita (right)*, and a colony of *Acropora (above)*.

The distance between the Australian coast and the Great Barrier is wider toward the south. At its northern end, the coral wall is close to the coast, shielding the continent from the high seas. At its southern end, from Townsville to Brisbane, the reef lies from 75 to 200 miles (120 to 320 kilometers) away from shore. Between the reef and the mainland is an interior sea known as the Great Australian Channel.

The structure of the reef also varies from north to south. In the south, where the water is colder, the coral wall is broken into a maze of secondary reefs, coral sandbanks, and atolls. Between these fragments of the great reef are broad channels deep enough to navigate. In the warmer waters farther north, however, the barrier grows wider, and navigable channels are few. In its northernmost 60 miles (100 kilometers), the reef is nearly uninterrupted, rising from the sea like a mountain range.

Most of the reef's surface remains underwater when the tide is high. Thus, the reef does little to defend the inner lagoon from the violence of the waves. On the windward side of the reef, facing the ocean, the water is white and foamy as it crashes against the coral. In these well-oxygenated waters, the greatest growth of polyps occurs. In contrast, on the leeward or inner side, the water is calm and navigable. From time immemorial, this side of the reef has been visited by whales. In this peaceful refuge they find a good place to bear their young.

Special Terms

cay: a small island usually of coral and covered with sand or vegetation
epithelium: the delicate tissue that covers a body or structure, or lines a cavity
equator: a line dividing the earth equally into the northern and southern hemispheres
exudate: a secretion that moistens an organism's outer surface
ichthyologist: a student of the branch of zoology dealing with fish
latitude: the distance from a point of the earth's surface to the equator, divided in degrees of meridian
leeward: the side of an island or land mass protected from the wind
masts and rigging: system of poles and ropes that support and control the sails of a ship
sextant: navigation instrument for verifying the latitude of a given point, using the sun to calculate its height and measuring the angle it forms with the horizon line
symbiosis: association of two living things, in which both benefit mutually
windward: side of an island or land mass in the path of the wind

Naked Coral
The soft corals in the order Alcionacea, such as those shown in the righthand photo, have skeletons that are not solid. These fleshy coral colonies are full of water. They reach a certain consistency thanks to a framework of fine limestone spines, and to the high concentration of salts that accumulate in their interior.

The zooxanthellae that usually accompany the reef-building coral need light in order to survive. For this reason, most living reef coral is found no more than 130 to 165 feet (40 to 50 meters) below the surface of the water, where the sun's rays can penetrate. But some coral walls in the Great Barrier Reef are hundreds of yards deep. At this depth, algae cannot live and coral cannot grow. What is the explanation of this mystery?

During World War II, the American navy discovered many submerged volcanic mountains, their crests strangely flattened, scattered through the Pacific Ocean. Various theories have been proposed to explain these structures, or *guyots*. One is that they had once been islands. Over time, the islands were worn down by the eroding force of the waves. Another explanation is that these were islands that were sinking slowly in the ocean.

A third theory is that the level of the ocean rose over time, submerging areas that had once been at the surface. This theory sheds the most light on the existence of deep-sea coral. Great changes in climate have occurred in the course of the earth's history. These range from Ice Ages to exceptionally hot periods. When the temperature in the polar regions rises, the icecaps melt, raising the level of the oceans.

This would explain why some coral walls can be found at depths where polyps could not now develop. The walls formed when the rocks that support them were very close to the surface. Later, new coral grew upon old coral to create reefs more than 3,000 feet (1,000 meters) thick.

Rich in oxygen and minerals taken from the coral, the waters surrounding the Great Barrier Reef are highly productive. They teem with plankton, a veritable "soup" of microscopic animals and plants that float in the upper layers of the ocean. The abundance of plankton attracts a diverse fish population.

Some 1,400 species of fish have been counted along the Great Barrier Reef. Most of these fish are highly colorful. Examples include the surgeon fish, the emperor, the anemone fish, the criss-cross butterfly fish, the scribbled angelfish, and the harlequin tusk fish. Others, such as the batfish, have less intense tones. Their coloration helps break up their silhouette to confuse predators.

There is only one classic shark species on the reef: the little pencil shark that lurks in coraline thickets, but is harmless to humans. The barracuda has a worse reputation. It is a superb swimmer, ferocious in appearance and immense in size, its mouth bristling with rows of sharp teeth.

In the Darkness

The dazzling shapes and colors in the world of the coral reef surpass the wildest fantasy. Among the most conservative designs is that of the brain coral, so named for its resemblance to the convolutions of the human brain *(opposite page, top)*. In contrast, the alcionarian *(bottom)* is exuberantly colorful. Oddly enough, it is usually found in the most poorly lighted parts of the coral structure—in caves or beneath great *Acropora*.

In the eighteenth century, an eminent French ichthyologist, Father Jean-Baptiste Labat, maintained that barracudas enjoy human flesh and guide themselves with their sense of smell. Labat reasoned that if the fish had to choose between a French and an English swimmer, it would inevitably devour the Englishman.

Labat reasoned that the less delicate feeding habits of the English, leaning toward strong alcoholic beverages, would send off a more powerful odor that would provoke the fish's attack. Perhaps the scientific reasoning of the past was less strict than it is today, but it was more charming.

The Great Barrier Reef is still almost untouched by humans. This is because it is so hard to reach and so dangerous for navigation.

During the 1960s, attempts to exploit the area for petroleum came under sharp criticism from Australian citizens, who were convinced of the need to preserve this valuable natural wonder.

So there is hope that the Great Barrier Reef may remain to flourish in its natural state. Coral will be able to continue its work for centuries to come, and our children and grandchildren will have the chance to enjoy one of the most beautiful scenes on the planet Earth.

The Challenge of Conservation

Few areas on the earth are degrading so rapidly as the marine environment. With the dumping of all kinds of products, commercial use for transport, and worldwide overfishing, our oceans are seriously threatened. The inclusion of certain privileged areas on the World Heritage list assures their conservation and points the way we must take if we want to maintain oceans worthy of the name.

Symbiosis

One of the most surprising adaptations in the hostile environment of the coral world is that of the clown fish. It lives among the stinging tentacles of the anemones, knowing that no safer place exists on the reef. To avoid being devoured by the anemone, it covers its body with a mucous substance that may prevent the anemone from firing its poisonous cells. In this way, the fish makes itself "invisible" to its otherwise deadly companion. On the other hand, the clown fish cleans the anemone's body and eats scraps of food the anemone leaves behind.

These Sites Are Part of the World Heritage

Aldabra Atoll: Located in the Seychelles Islands, this is a group of four coral islands that enclose a lagoon. Around the outside, they are well protected by a coral reef. The atoll shelters the world's greatest population of giant turtles. It is also a refuge for green turtles and Carey turtles and has important colonies of seabirds.

Great Barrier Reef: Along the eastern coast of Australia, the Great Barrier Reef is the most extensive coral zone on earth—1,250 miles (2,000 kilometers) in length, 100,000 miles (260,000 square kilometers) in area, and encompassing more than 2,500 distinct reefs. It harbors a variety of marine fauna, including over 1,400 species of fish and over 250 species of coral. The reef also serves as refuge for endangered species, such as the great green turtle.

Glossary

algae: aquatic plants, some as small as one cell, that manufacture their own food through photosynthesis

atoll: a coral reef surrounding a lagoon

carnivore: a creature that eats animals

corvette: a swift-moving sailing vessel

extinction: the complete dying out of a species of animal or plant

herbivore: a plant-eating animal

invertebrate: an animal that does not have a spinal column

lagoon: a shallow body of water near or connected to a larger body

mangrove: a tree with large, exposed roots that grows on banks or in marshy areas

photosynthesis: the process by which plants take in carbon dioxide and energy from the sun and release oxygen and other substances

phylum: one of the primary classifications of creatures in the animal world

plankton: tiny plants and animals that float in a body of water

reptile: an animal that moves along on its belly or on very short legs, such as snakes, lizards, and turtles

salinity: the degree of saltiness of a body of water

sediment: matter that settles to the bottom of a body of water

sextant: an instrument that sailors use in navigating

silhouette: a dark form or shape, as seen against a lighter background

submerged: underwater

substratum: an underlying base or foundation

Index

Page numbers in boldface type indicate illustrations.

Titles in the World Heritage Series

The Land of the Pharaohs
The Chinese Empire
Ancient Greece
Prehistoric Rock Art
The Roman Empire
Mayan Civilization
Tropical Rain Forests of Central America
Inca Civilization
Prehistoric Stone Monuments
Romanesque Art and Architecture
Great Animal Refuges
Coral Reefs

Photo Credits

Front Cover: F. Candela/Incafo; p. 3: F. Candela & H. Geiger/Incafo; p. 4: F. Candela & H. Geiger/Incafo; p. 5: H. Geiger/Incafo, F. Candela & H. Geiger/Incafo; p. 7: F. Candela & H. Geiger/Incafo; p. 8: F. Candela & H. Geiger/Incafo; p. 9: F. Candela & H. Geiger/Incafo; p. 11: F. Candela & H. Geiger/Incafo; pp. 12-13: F. Candela & H. Geiger/Incafo; p. 15: F. Candela & H. Geiger/Incafo, P. Roberts, P. Roberts; p. 17: F. Candela & H. Geiger/Incafo; p. 18: F. Candela & H. Geiger/Incafo; p. 19: P. Roberts, P. Roberts; pp. 20-21: F. Candela & H. Geiger/Incafo; p. 23: J. & J. Blassi/Incafo; p. 24: F. Candela & H. Geiger/Incafo; p. 25: F. Candela & H. Geiger/Incafo, F. Candela/Incafo; p. 27: F. Candela & H. Geiger/Incafo; p. 29: F. Candela/Incafo; F. Candela & H. Geiger/Incafo; p. 30: F. Candela & H. Geiger/Incafo; p. 31: P. Roberts, F. Candela & H. Geiger/Incafo, F. Candela & H. Geiger/Incafo; back cover: F. Candela & H. Geiger/Incafo, F. Candela/Incafo.

Project Editor, Childrens Press: Ann Heinrichs
Original Text: Alberto Ruiz de Larramendi
Subject Consultant: Susan Gray
Translator: Deborah Kent
Design: Alberto Caffaratto
Cartography: Modesto Arregui
Phototypesetting: Publishers Typesetters, Inc.

UNESCO's World Heritage

The United Nations Educational, Scientific, and Cultural Organization (UNESCO) was founded in 1946. Its purpose is to contribute to world peace by promoting cooperation among nations through education, science, and culture. UNESCO believes that such cooperation leads to universal respect for justice, for the rule of law, and for the basic human rights of all people.

UNESCO's many activities include, for example, combatting illiteracy, developing water resources, educating people on the environment, and promoting human rights.

In 1972, UNESCO established its World Heritage Convention. With members from over 100 nations, this international body works to protect cultural and natural wonders throughout the world. These include significant monuments, archaeological sites, geological formations, and natural landscapes. Such treasures, the Convention believes, are part of a World Heritage that belongs to all people. Thus, their preservation is important to us all.

Specialists on the World Heritage Committee have targeted over 300 sites for preservation. Through technical and financial aid, the international community restores, protects, and preserves these sites for future generations.

Volumes in the *World Heritage* series feature spectacular color photographs of various World Heritage sites and explain their historical, cultural, and scientific importance.